Shrinking Biodiversity. Current Status and Perspectives in Madagascar and Tanzania

Elnur Aliyev

Bibliographic information published by the German National Library:

The German National Library lists this publication in the National Bibliography; detailed bibliographic data are available on the Internet at http://dnb.dnb.de.

ISBN: 9783346481825
This book is also available as an ebook.

Nymphenburger Straße 86
80636 München

Print and binding: Books on Demand GmbH, Norderstedt, Germany
Printed on acid-free paper from responsible sources.

GRIN web shop: https://www.grin.com/document/1117693

2021

SHRINKING BIODIVERSITY - CURRENT STATUS AND PERSPECTIVES IN MADAGASCAR AND TANZANIA

Luisa Desch and Elnur Aliyev
Rhine-Waal University of Applied Sciences
7/7/2021

Table of Contents

List of Abbreviations

CBD	Convention on Biological Diversity
CDC	Country Development Cooperation
COVID-19	Coronavirus disease
DLC	Duke Lemur Center
DNA	deoxyribonucleic acid
FAO	The Food and Agriculture Organization
GDP	Gross Domestic Product
IUCN	International Union for Conservation of Nature
MAST	Manifattura di Arti, Sperimentazione e Tecnologia
MEA	Millennium Ecosystem Assessment
MEF	Ministry of Environment and Forests
MRPA	Managed Resources Protected Areas
NBSAP	National Biodiversity Strategies and Action Plans
NGOs	Non-Governmental Organizations
PA	Protected Area
PROTECT	Promoting Tanzania's Environment, Conservation, and Tourism
RTI	Research Triangle Institute
SCBD	The Secretariat of the Convention on Biological Diversity
UNDP	The United Nations Development Program
UNEP-WCMC	The UN Environment Programme World Conservation Monitoring Centre
UNESCO	The United Nations Educational, Scientific and Cultural Organization
USAID	The United States Agency for International Development
USD	The United States Dollars
WCS	Wildlife Conservation Society
WD-OECM	World Database on Other Effective Area-based Conservation Measures
WDPA	The World Database on Protected Areas
WWF	World Wildlife Fund

Abstract

In the post-industrial period and currently, the biodiversity loss is accelerating globally. Tanzania and Madagascar possess a very high degree of biodiversity of ecosystems, plants and wildlife sustaining critically the livelihoods and wellbeing of millions of people. Both countries face extinction of endemic, as well as other species, degradation of habitats as a result of mainly the poverty, climate change, invasive species, agricultural expansion and lack of effective natural resources management and conservation education of population. In the framework of Convention on Biological Diversity, these East African countries have set goals and targets to tackle the imminent and long-term threats to biodiversity. However, alarming trends continue and threaten the ecosystems and vulnerable species of the countries with extinction such as deforestation and species extinction among others. More investment needs to be spent on existing research and effective management practices if the loss of biodiversity is to be reversed.

"We have reached an unprecedented moment whereby humans change the Earth and its processes more than all other natural forces combined".

Edward Burtynsky - The Anthropocene Project, MAST Foundation

In almost all Earth's ecosystems, biodiversity loss is on the rise. In order to grasp the seriousness of this global trend it is necessary to have a common understanding of the concept and know why it is important at all. The Convention on Biological Diversity (CBD) defines biological diversity as being "the variability among living organisms from all sources including, inter alia, terrestrial, marine and other aquatic ecosystems and the ecological complexes of which they are part; this includes diversity within species, between species and of ecosystems" (Convention on Biological Diversity, 1992). The definition delineates three aspects of biodiversity, namely, genetic diversity, species diversity and ecosystem diversity. Genetic diversity is based on the idea that even the members of same species are different at genetic or DNA level and this drives the evolution process by natural selection (Hughes, Brian D.Inouye, Marc T. J. Johnson, Nora Underwood, & Mark Vellend, 2008). Species diversity is a simple measure of count of species within a genus (Macarthur, 1965), while ecosystem diversity refers to quantity, types and patterns of terrestrial and marine/wetland ecosystems and the living processes in them (Burton V. Barnes & Marc Lapin, 1995).

The importance of biodiversity is not complex to understand. Considering the values of ecosystems and species diversities such as environmental services that sustain different crucial life cycle processes, their economic value and socio-cultural role for especially the developing countries, it is apparent that even one species matters when it comes to wellbeing of nature, including the humans. However, unfortunately, in the post-industrial period and currently, the biodiversity loss is accelerating. A feature article published in 2009 in *Nature* journal illustrates the nine planetary boundaries and shows that biodiversity boundary has been critically trespassed thanks to anthropogenic impacts threatening the functioning of the whole planet (Rockström, et al., 2009). Although at a preliminary level, this article proposes the extinction rate as an indicator of when Earth would lose its resilience due to biodiversity loss and suggests a tipping point of ten species per million species per year. It also states that this century would witness climate change as the main drivers of biodiversity loss coupled with other human-induced changes leading to almost 30% of mammals, amphibians and bird species being in danger of extinction.

The current paper discusses the biodiversity status and perspectives of two of biologically and culturally richest countries in the world – Tanzania and Madagascar. These East African countries possess a very high degree of biodiversity of ecosystems, plants and wildlife sustaining critically the livelihoods and wellbeing of millions of people. At same time, Tanzania and Madagascar have one of lowest human development indices in the world emphasizing the value of biodiversity to these countries and explaining the current rate of overexploitation and other human activities in relation to biodiversity.

Tanzania

In this section, biodiversity of Tanzania is discussed in more detail. The landscape and significance of biodiversity in Tanzania will be covered first, followed by the present state of biodiversity and threats. Conservation programs and a section conclusion will be addressed at the end of this section.

Value of Biodiversity to Tanzania

Tanzania is East Africa's largest country, with the islands of Zanzibar, Pemba, and Mafia. Besides the modest coastal belt of the mainland and the surrounding islands, the majority of mainland Tanzania is over 200 meters in height. The East African Rift System cuts through mainland Tanzania in two north-south sections, creating a span of narrow, deep depressions mainly filled with lakes (National Geographic, n.d.). Between the two branches is the central plateau, which covers more than a third of the nation. Africa's highest point, Mount Kilimanjaro, is surrounded by three of the continent's greatest lakes: Lake Tanganyika (world's second deepest lake), Lake Nyasa and the world´s second-largest freshwater lake, Lake Victoria (National Geographic, n.d.). Tanzania's inland water covers roughly 59,000 square kilometres, due to its many lakes. The diversity of soils in mainland Tanzania outnumbers that of any other African country. The most fertile soils are reddish brown volcanic soils found in the highlands. Many river basins offer good soils as well, although they are prone to floods and require drainage management. The inner plateaus' red and yellow tropical loams, however, have a moderate-to-poor fertility (Mascarenhas et al., n.d.).

Tanzania's mainland is divided into four distinct climatic and geographical zones:" the hot and humid coastal lowlands of the Indian Ocean shoreline, the hot and arid zone of the broad central plateau, the high inland mountain and lake region of the northern border, where Mount Kilimanjaro is situated, and the highlands of the northeast and southwest, the climates of which range from tropical to temperate" (Mascarenhas et al., refers to climate section., n.d.).

Approximately 90 percent of Tanzanians live in rural regions and rely on what they can cultivate on their land to survive. Tanzania has a significant number of natural resources that boost the economy and improve human well-being. Hence, biodiversity is a big driver for Tanzania's economy. Given the above, Tanzania´s wealth of biodiversity is extremely important to the country. Fishing, livestock, agriculture, and forestry combined account for more than 65 percent of GDP, more than 60 percent of total export earnings and over 80 percent of total employment (5th National Report, 2014). The country is benefiting considerably from biodiversity as it is crucial for attracting tourists, food source, pharmaceuticals, for decomposing organic waste and soil conditions, and building energy and materials. Likewise, the industry is relying heavily on biodiversity especially agro, forest, pharmaceutical- and food industries. Tanzania is one of the world's greatest reservoirs of biodiversity due to its huge national parks, wetlands, marine, and freshwater systems, significant reservoirs of plant and animal species, the Eastern Arc mountains, and coastal forests. Tanzania has 18 endemic lizard species, ten bird species, 31 endemic amphibian species, nine snake species, roughly 80 percent of the renowned African violet flowers, and 40 percent of the world´s wild coffee types (5th National Report, 2014). Moreover, World Heritage Sites add a great value to the biodiversity: Serengeti National Park, Selous Game Reserve, Kilimanjaro National Park, Ngorongoro Conservation Area as well as three Biosphere Reserves, i.e. Lake Manyara, Ngorongoro – Serengeti and East Usambara. These reserves are home to a diverse range of mammals, reptiles, birds, and amphibians, all of which contribute significantly to security of food, government revenue, and community income (5th National Report, 2014).

Biodiversity Status

The United Republic of Tanzania is very rich in biodiversity and one of the mega-biodiversity rich countries worldwide. Tanzania's mega-biodiversity is found in both protected and non-protected places and encompasses ecosystems, species, and genetic resources. "Tanzania's natural ecosystems are split into three groups: terrestrial ecosystems, coastal and marine ecosystems, and inland water ecosystems (lakes, rivers, dams, and wetlands)" (NBSAP, 2015, p. 11).

Tanzania hosts six of the 25 globally known biodiversity hotspots: The Great Lakes for Cichlid fishes; the Marine Coral Reef ecosystems; The Eastern Arc Old Block-Mountain Forests; the Coastal Forests; the Ecosystems of the Alkaline Rift-Valley Lakes; and the grassland savannas for large mammals (e.g., Serengeti Park) (SCBD, n.d.). The Eastern Arc Mountains in Tanzania and Kenya comprise the majority of Tanzania's wet montane tropical forest and are East Africa's biologically richest area for its size. They also have the highest percentage of endemic species of any East African region and are one of the world's 17 most endangered tropical forest ecosystems or global hotspots (Newmark, 2002).

Terrestrial Ecosystems

Forests, mountains, drylands, savannah, and agricultural lands are types of terrestrial ecosystems. Tanzania's forest cover is approximately 48 million hectares (about 55 percent of the total land area), with woods accounting for around 51 percent of the total land area, or 93 percent of the forest area. The rest of seven percent are lowland forests, mangrove forests, plantations, and humid montane forests. Additionally, Tanzania has roughly 44 million hectares of agricultural land (NBSAP, 2015).

The natural forests of Tanzania are divided into three categories: montane forests, mangroves and miombo woods. The majority of the montane forests have a high-water catchment value, making them a key river source. Yet, African Sandalwood field stocks have plummeted, and human pressure on this species is so intense that it is on the verge of extinction unless immediate silvicultural treatments[1] are implemented (6th National Report, 2019). Mangroves are high-productivity environments that produce significant amounts of organic matter that provides food for a variety of organisms. They further contribute to the stabilization of the shoreline by preventing erosion, filtering water, and settling sediments that might otherwise harm seagrass habitats and coral reefs. However, the density of Mangroves has decreased significantly over the years (Torell et al, 2007).

Species and Plant Diversity

The country has a large number of species, with at least 14,500 known and documented species, and is one of the 15 countries in the world with the most endemic and vulnerable species. It accounts for nearly four percent of all threatened species worldwide. Tanzania is also home to more than a third of Africa's total plant species and ranks 12th in the world in terms of bird species (SCBD, n.d.). Tanzania has between 400 and 3,000 endemic species, with mammals and cycads having the highest proportion of threatened species, and amphibians and plants having the highest number of threatened endemic species. Endemic plants are African Blackwood (Mpingo), African Teak (mninga), Erythrina schliebenii, Karomia gigas (one of the world's rarest trees) and

[1] Silvicultural treatments are applied to trees and stands to change, accelerate, or maintain their condition (https://www.nrs.fs.fed.us/fmg/nfmg/fm101/silv/p2_treatment.html, as of 03.07.2021)

Uvariodendron gorgonis (evergreen tree) and marine animals as sea turtles, as well as coelacanth and dugongs (NBSAP, 2015). Over the previous decade, the number of vulnerable species in the country has nearly tripled, which can be attributed to habitat loss, overexploitation, fragmentation, and degradation, along with climate change effects (SCBD, n.d.). The black rhinoceros, shoebill stork, wild dog, Kihansi Spray toad, chimpanzee, red colobus monkey, African elephant, cheetah, and wattled crane all have significant populations in Tanzania that are endangered and threatened. The savannah grasslands, defined by the dry miombo woods, make up a large portion of the wildlife region (SCBD, n.d.). Tanzania has a high level of species endemism that can be linked to the country's complex topography and biological isolation in some places, generating unique microclimates and ecological circumstances that support the several endemic species (NBSAP, 2015).

Coastal and Marine Ecosystem

The coastal zone of Tanzania is rich in wildlife and natural resources. Coral reefs, mangrove forests, seagrass beds, and maritime fisheries are all significant ecosystems. Tanzania's coastal and marine ecosystems cover roughly 241,500 km^2, which is about 20 percent of the total land area of the country. Coral reefs are high-diversity, high-productivity tropical shallow-water habitats. They run the length of Tanzania's continental shelf, which is 600 kilometres long. Tanzania's coral reefs are home to approximately 500 species of fish and other invertebrates, making them a valuable fisheries resource that supports 90 percent of artisanal marine fisheries over 3,580 km^2. However, human, and natural threats have decreased coral reefs tremendously (Torell et al, 2007).

Inland Water Ecosystem

Lakes, rivers, springs, natural ponds, subterranean sources, man-made reservoirs, and wetlands are among Tanzania's abundant freshwater resources. Lakes cover roughly six percent of the land area and include the transboundary major lakes that are Lake Tanganyika, Lake Victoria and Lake Nyasa. *Kilombero*, *Rufiji*, *Wami*, *Mara*, *Ruaha*, *Ruvuma*, *Malagarasi*, *Kagera*, and *Pangani*, as well as their tributaries and accompanying small streams, form a varied network of permanent and seasonal rivers. Many rivers are not protected, with the exception of a few rivers located within protected areas, and hence face decreasing ecological integrity as well as disturbance of ecosystem goods and services they provide (NBSAP, 2015). The large lake system, inland drainage systems, important river networks, and deltaic mangroves account for around ten percent of the country's territory that equals around 88,300 km^2 of total land area. Wetlands are one of the most productive ecosystems and are consequently essential for energy generation, water retention, flood control, groundwater recharge, and river and lake eutrophication prevention, as well as, maintaining specialized biota and traditional uses. Many wetlands were automatically protected a few decades ago because of their remoteness, size, and negligible utility for agriculture or other economic activity. Nonetheless, as a consequence of various socioeconomic developments, the country's wetlands have recently undergone fast conversion, leading to their degradation. Tanzania's wetlands are home to about 650 different species of molluscs, crabs, and fish (6th National Report, 2019).

Threats to Biodiversity

The Millennium Ecosystem Assessment report demonstrates that biodiversity loss is related to environmental changes that have occurred at a faster rate in the last 50 years than at any other time in human history (MEA, 2005). Tanzania's biodiversity is facing significant reductions in ecosystem quality, species abundance, and diversity, despite its biological richness and conservation efforts. There are many pressing issues confronting biodiversity in Tanzania. The most urgent are: resource overexploitation, overgrazing, over-dependency of fuel wood for domestic use and bush fires as well as other influences on biodiversity and conservation as poverty, a lack of alternative energy sources, influx of immigrants, spread of alien species (*water hyacinth, Nile perch*), and a lack of community awareness about biodiversity protection (6th National Report, 2019). Invasive alien species are one of the world's most serious dangers to indigenous biodiversity, and their invasion and persistence will eventually result in a drastic reduction in biodiversity. They also have the potential to outgrow indigenous species. Over 60 species have been reported as invasive species in Tanzania (Kideghesho, Mwamende & Selemani, 2013). Agricultural methods, changes in atmospheric composition and climate change, and biological pest management are the biggest factors through which invasive alien species are introduced (NBSAP, 2015).

In addition, bush fires are on the rise, burning an average of eleven million hectares per year, primarily as a result of human activities such as farm preparation, honey harvesting, game hunting, pasture improvement burning, and charcoal burning (NBSAP, 2015). Other threats are illicit fishing, deforestation, illegal hunting and logging, mineral and aggregate mining, unplanned human settlement developments, agriculture, grazing and livestock migration, water quality degradation and cultural beliefs (6th National Report, 2019). Especially poaching and unrestricted legal hunting have resulted in overexploitation of wildlife. (Kideghesho, Mwamende & Selemani,

2013). Due to Tanzania's distinctive species and fairly well-developed transportation infrastructure, it has emerged as a key destination for wildlife trafficking. As a result, well-armed poachers and well-organized trafficking networks have annihilated populations of rhinos, pangolins, big cats and elephants.

Moreover, local villages are pushed to over exploit their resources for living, as natural resource management strategies for livelihoods are generally unsustainable and not efficient (USAID, 2001). In many parts of the country, agricultural growth is a major contributor to habitat fragmentation and biodiversity loss; there is a growing demand for grazing areas and livestock feed as well as crop production. The number of goats and cattle grew tremendously in recent years. The expansion of agricultural and grazing land, along with unsustainable agricultural methods, has resulted in fragmentation of natural ecosystems, increasing biodiversity challenges (NBSAP, 2015).

Further, the decline of marine and coastal living resources, the devastation of coral reefs, coastal contamination, and erosion are the primary challenges to marine and coastal ecosystems. Destructive fishing activities account for much of the destruction of reef ecosystems. The usage of dynamite, which has been used in some areas of Tanzania for almost 40 years, is the most harmful fishing method. Every explosion instantaneously kills all fish and most other living organisms within a 15-to-20-meter radius, as well as entirely destroying the reef ecosystem (6th National Report, 2019). The destruction to coral reef structures is severe and mostly irreversible. Nevertheless, dynamite fishing has decreased across the country as a consequence of public awareness, increased regulation enforcement, and the formation of a Multi-Agency Task Team, which combats environmental and wildlife crime, among other things. However, small mesh seine nets used to catch fish on the bottom and around reefs are nearly as destructive as explosives (6th National Report, 2019). A lack of employment opportunities and education, poverty and rising demand for shellfish and fish work against incentives for using more sustainable fishing practices (Torell et al., 2007).

Additionally, over-harvesting of mangrove for firewood, construction poles, charcoal, and boat construction, which account for about 46 percent of mangrove forest degradation, while clear-cutting of mangrove for agriculture, hotel building, urbanization, solar salt production, and road construction account for another 30 percent. Rising demand and prices for mangrove charcoal and firewood from Zanzibar and Dar es Salaam will encourage even more people to participate in mangrove harvesting (Torell et al., 2007).

Furthermore, climate change, pests and diseases, and incorrect agro-chemical use harm pollinators and other important non-target organisms, putting biodiversity in the agricultural ecosystem at jeopardy. Climate change has an impact on nearly all ecosystems in Tanzania. Likewise, improper disposal of industrial and household waste on land and water bodies are polluting surface and groundwater supplies (6th National Report, 2019). The increasing reliance on artificial fertilizers and chemical sprays for farming, also close to rivers, is another significant issue (Torell et al., 2007). Substantial amounts of nutrients, mostly phosphates and nitrates, are present in most aquatic ecosystems as a result of household, industrial, and agricultural activities. Toxic substances from industries and mining activities, agricultural herbicides and pesticides, and sewage-effluent find their way into aquatic systems, and the majority of them are hazardous to wildlife and can cause long term damage (NBSAP, 2015).

Population growth is another factor for biodiversity loss. Tanzania's population climbed from 12.3 million in 1967 to 44.9 million in 2012 and is expected to reach 59.8 million by 2025. The majority of people in rural areas rely on subsistence agriculture for food and money, which leads to unsustainable resource extraction and habitat degradation, culminating in biodiversity loss (NBSAP, 2015).

Trends in Biodiversity

Tanzania has been a member of the Convention on Biological Diversity since 1996, and it is committed to upholding its international commitment to safeguard and maintain its biodiversity as a global resource. It is essential that the country protects its biodiversity, and it has taken a variety of steps to do so. The updated National Biodiversity Strategy and Action Plan (NBSAP) (2015-2020) of the government aims to address national targets built on objectives that contribute to global goals. Tanzania also acknowledges the need to develop new policies in response to new and emerging policy challenges, as well as the need to assess and improve the execution of current policies, plans, and strategies (SCBD, n.d.). About 40 percent of the total terrestrial land is already under protection. According to the 2009 nationwide elephant census, elephant populations in protected regions like Tarangire-Manyara and Serengeti increased by 64 percent, correspondingly. Nature parks and reserves have been established for the purpose of developing and administering nature reserves in Zanzibar. The Chumbe Island Coral Park Reef Sanctuary has been designated as one of the most diversified regions, with 90 percent of East Africa's hard coral species estimated to be found there (NBSAP, 2015). Several antelopes of the severely endangered Ader's duiker antelope and a huge population of the Coconut crab, as well as numerous endangered bird species, can be found in the Closed Forest Habitat. The East Africa

Marine Ecoregion recognizes Jozani National Park as home to at least 291 vascular plant species from 83 families, 28 of which are indigenous, and 21 species that are vulnerable or endangered. The national park is also home to a diverse range of animals, both terrestrial and marine species, many of which are rare and endangered, including the Zanzibar leopard and the Zanzibar red Colobus monkey (NBSAP, 2015).

Given the country's backward biodiversity trends, more action is necessary. Despite having protected areas, Tanzania has lost at least one-third of its key ecosystems in the previous several decades (SCBD, n.d.). Forests and other types of woodlands are disappearing, while land utilized for commercial and urban developments is expanding. In comparison to other agro-ecological zones, the southern highlands, plateau, and semi-arid agro-ecological zones have seen substantially greater levels of degradation. Poor agricultural practices and overgrazing are mostly to blame. As Tanzania's population grows, so does the need for food, resulting in increasing habitat loss and strain on the biodiversity (NBSAP, 2015).

In addition, Tanzania's mainland has lost around 18 percent of its forest cover, roughly at a rate of 0.7 percent each year. In Zanzibar, almost half of the mangrove forests are deemed degraded. Almost 90 percent of wetlands are under threat, with many of their functions being lost, and almost half of inland water habitats have been damaged (NBSAP, 2015). A substantial number of species are also on the decline in terms of species diversity. When compared to the numbers recorded in the year 2000, the proportion of endangered species in the country has nearly tripled. This is due to a variety of reasons, including fragmentation, effects of climate change, and habitat loss, as well as degradation. Moreover, the elephant population is falling, and a 60 percent decline has been noted. Black rhinoceros' numbers are also on the decline. Overfishing has resulted in a decrease of prawns and sea turtles, and dugongs are becoming extremely scarce. Due to a variety of reasons, especially invasive alien plants, the Lake Victoria Basin´s estimated 200 indigenous fish species have decreased (NBSAP, 2015).

Nature Conservation Programs

A variety of regulations and legislation exist in Tanzania that impact biodiversity conservation and management policies. There are also numerous initiatives and programs that try to conserve biodiversity. Some will be addressed in this paragraph.

There is an extensive system of protected areas from grasslands to Africa's highest mountain comprising four natural World Heritage Sites: Kilimanjaro National Park, Ngorongoro Conservation Area, Serengeti National Park, and Selous Game Reserve. Mafia Island National

Park, Tanzania's first legal marine protected area, was established in 1995. Since then, the number of protected marine habitats has increased (IUCN, 2017). The country has also dedicated itself to highly ambitious national conservation goals, as well as the Convention on Biological Diversity's Strategic Plan for Biodiversity 2011-2020. Tanzania's attempts to accomplish these ambitious targets in terrestrial and inland waterways, as well as, coastal and marine habitats, are guided by Aichi[2] Target 11, including both quantitative and qualitative features (IUCN, 2017).

Tanzania has established a vast network of forest and wildlife protected areas, encompassing one Conservation Area, 16 National Parks, 109 Forest Reserves, 42 Game Controlled Areas, 28 Game Reserves, four Ramsar Sites and 38 Wildlife Management Areas. UNESCO has designated four protected places as World Heritage Sites, three of which are Biosphere Reserves. Lake Manyara, Ngorongoro - Serengeti, and East Usambara are the Biosphere Reserves. Kilimanjaro National Park, Ngorongoro Conservation Area, Selous Game Reserve, and Serengeti National Park are all World Heritage Sites in the Nature category (NBSAP, 2015).

There are 21 Marine Protected Areas that consist of 17 Marine Reserves and four Marine Parks. Only 2,173 km^2 (about 6.5%) of the territorial sea of Tanzania Mainland have been demarcated as Marine Protected Areas. *Ruvuma Estuary Marine Park, Mnazi Bay, Mafia Island Marine Park*, and *Tanga Coelacanth Marine Park*, as well as 15 Marine Reserves, are the protected areas in Tanzania's mainland. There are two Marine Reserves and one Marine Park in Zanzibar (NBSAP, 2015).

The Promoting Tanzania's Environment, Conservation, and Tourism (PROTECT) initiative, coordinated by the Research Triangle Institute (RTI), was established by the United States Agency for International Development (USAID) in 2015. The USAID PROTECT program, which runs for five years, aims to develop institutions, and create skills for Tanzanians to better protect their biodiversity while also safeguarding their economic growth. That is why, the private sector plays a crucial part as an important conservation partner (RTI, n.d.). The partnership operator Asilia Camps and Lodges has a strong focus on conserving cheetah breeding habitats in the Eastern Serengeti. The project supplies crucial patrol equipment to the Tanzania National Parks Authority unit for anti-poaching. In the Pemba Channel Conservation Area, USAID PROTECT collaborated with the Manta Resort's Kwanini Foundation to restore and preserve coral reefs. This region is an important marine biodiversity hotspot that is exceptionally resistant

[2] The Aichi goals are statements of worldwide biodiversity conservation objectives. They were approved in Nagoya, Aichi Prefecture, in 2010 at the 10th Conference of the Parties to the Convention on Biological Diversity. By 2020, the objectives should be met (IUCN, 2017).

to coral bleaching, yet it was on the verge of extinction mainly due to dynamite fishing and other harmful human activities (RTI, n.d.). Other crucial goals of the USAID PROTECT: fighting wildlife crime, strengthening local institutions' capacity, supporting land and sea conservation activities, improving Tanzania´s conservation policies and regulations, and developing a long-term environment for conservation work (RTI, n.d.).

The Wildlife Conservation Society (WCS) has worked in Tanzania since 1956, with a single nation program in place since 2006. WCS has undertaken a wide range of initiatives, including training, research, monitoring, institutional and community assistance, education, and the establishment, expansion, and maintenance of major protected areas. WCS creates community-based projects to support improved management of essential species and ecosystems, ensuring their long-term preservation and integrity. Communities in the vicinity of protected areas are provided assistance in natural resource management so that they may directly benefit from conservation, safeguard water and fuel supplies, and effectively manage conflicts between human and animals. WCS assists government and non-governmental organizations as well in managing and monitoring significant landscapes and species around the country (WCS Tanzania, 2016).

There is also the Eastern Arc Mountain Conservation Program, which started in 2003. The goal is to create and execute conservation measures that will assure the long-term protection of the Eastern Arc Mountain forests, both in terms of forest conservation and biodiversity (NBSAP, 2015).

To preserve the coral reefs, the Tanzania Coral Reef Task Force was established in 2020 with the goal to improve local capacity in research, governance, and management and provide a forum at national level for sharing information on regional projects. It also intends to examine, manage, and coordinate legal elements of the country's efforts to reduce dynamite fishing. Every two years, they provide a Coral Reef Status Report on current issues and situations concerning the coral reefs (NBSAP, 2015).

The Environmental Act of 2004 establishes a legislative and institutional framework for environmental sustainability, pollution prevention and control, waste management, environmental quality standards, public engagement, compliance, and enforcement (Torell et al., 2007).

Section Conclusion

In summary of the Tanzania section, it seems that the government is already setting policies and measures to protect biodiversity, nevertheless it needs to analyse and monitor invasive species as well as create effective management methods. The government is already working on conserving biodiversity and has implemented some policies and programs like the Water Policy (2002), Forest Act (2002), Wildlife Conservation Act (2003), Fisheries Act (2002) and Environmental Act Plan (2004). In the National Biodiversity Strategy and Action Plan they have set goals and targets for biodiversity conservation.

The government's vision overall is to have biodiversity protected and restored by 2025 (NBSAP, 2015). But, apart from national goals, people need to be addressed and educated. For communities and rural people, biodiversity is essential for food, medicine, and a place of living. At the same time, due to their practices, they also destroy biodiversity. That is why they should be included in conservation efforts, decision-making and planning, as well as development-related actions (Newmark, 2002). Local communities will be encouraged to match their behaviour with conservation aims if they see biodiversity resources, such as animals, as an asset instead of a burden. Conservation education should be addressed in both urban and rural populations. Conversation plans often fail due to insufficient understanding of the significance of biodiversity in ecosystem and human health by affected parties, and a lack in financial resources (Newmark, 2002). People become more aware of the value of conservation and biodiversity as a result of increased education and awareness, and they take action to conserve biodiversity.

Poverty is one of the primary drivers of biodiversity loss, and it needs to be addressed. Especially communities that damage biodiversity in order to survive should be supplied with better livelihood alternatives generating opportunities. Climate change is a key driver of biodiversity loss and necessitates the implementation of mitigation and adaptation strategies. Such strategies should focus on limiting or regulating human activities like deforestation, implementing good land management, greater reliance on alternative energy technologies, and guaranteeing adequate fire management (Newmark, 2002).

Biodiversity is a vital indication of a healthy ecosystem. We are all connected, and the extinction of a species can have unanticipated consequences, sometimes leading to the annihilation of entire ecosystems. As a result, it is critical to protect biodiversity and, by extension, the lives of all who live on this world.

Madagascar

Madagascar is one of the 36 biodiversity hotspots with a possession of rare ecosystems that accommodate a substantial diversity of flora and fauna (Explore the Biodiversity Hotspots, 2021). The importance of biodiversity is amplified by existing high proportion of endemic species in natural habitats of the island. According to World Wildlife Fund (WWF, 2021), about 95% of reptiles, 89% of plants and 92% of mammals that emerged in Madagascar throughout millions of years are found in no other place in the world.

However, several factors such as illegal wildlife trafficking, unsustainable use of natural resources including deforestation, weak governance and corruption, emergence of threatening alien species and climate change are causing the shrinkage of biodiversity at an alarmingly fast rate. These regressive evolution is disastrous for the rich natural habitat as well as the social and economic survival and development of the Malagasy people residing in the island. In order to understand the extent of biodiversity loss and its implications, the current chapter discusses the multiple values of biodiversity to Madagascar. Then following the current status and ecological trends, threats to the ancient biodiversity are mentioned. The last section of the chapter reviews the nature conservation programs and projects in place by a plethora of international and national stakeholders and considers the perspectives of biodiversity in Madagascar.

Value of Biodiversity to Madagascar

The vast amount of wealth that Madagascar possesses in the form of biodiversity plays a substantial role in the sustainable development of economic, environmental, socio-cultural and scientific status of Malagasy people. Well-being of a large portion of local population depends, to a great degree, on several biodiversity elements such as flora and fauna abundance, many undeveloped areas free from anthropogenic impact and diversity of terrestrial and aquatic ecosystems among others.

Economic Values

One level on which Madagascar's flora and fauna contribute to the country economy is the international trade (MEF, 2014). The available data from the 5th National Report of Madagascar show that income from the export of forest products accounted for 0.02 % of total country exports in 2011 (Table 1). It can be seen that from 2009 to 2012 the most revenue-generating forest products were essential oils and medicinal plants which accounted for more than

70% of total forest export incomes throughout the period. These medicinal and aromatic plants mainly constitute valuable raw materials for manufacturing industries. They were followed by different kinds of woods (on average 13%) and manufactured goods (on average 7.5%).

Table 1. Forest Products Exports Income (in US $)

Type of products	2009	2010	2011	2012
Pinewood	202.606,30	884.039,68	727.430,75	137.222,01
Rosewood	91.426,26	59.388,48	26.255,91	1.749,38
Plain wood	461,62	12.924,88	-	
Ordinary woods	37.419,93	85.629,45	32.281,46	30.349,72
Sub-total woods	*331.914,11*	*1.041.982,50*	*785.968,12*	*169.321,11*
Manufactured products	526.018,81	401.033,31	219.999,38	260.759,08
Secondary products	20.344,62	14.756,16	8.066,57	8.457,50
Essential oils and medicinal plants	4.634.515,05	4.068.421,50	3.656.370,21	2.408.360,05
Flora and fauna	152.067,49	157.430,21	109.888,91	143.871,03
TOTAL	**5.664.860,08**	**5.683.623,68**	**4.780.293,20**	**2.990.768,77**
% of total exports			**0,02**	

Source: (MEF, 2014), Own illustration

In comparison to forest products, biodiversity in the form of fish and agricultural products and by-products plays substantially more role in international trade of the country (0.68% of total exports in 2011). Table 2 summarizes the different contributions by categories for the same four-year period. As can be seen, crops specifically grown for their commercial value (cash crops) account for almost half of agricultural export value in all years, while produces of plant origin bring in slightly less money and is followed by food and animal origin (by-)products. In total, fishery products take up a share of around 40 million USD mainly represented by shrimp and various types of crustaceans/molluscs exports.

Table 2. Other Products Exported by Madagascar

Type of products	2009	2010	2011	2012
Fishery Products				
Fish	1.494.373,53	2.082.631,61	2.223.739,56	
Various Shellfish	44.858,03	70.269,79	70.110,42	
Shrimps	38.326.676,12	30.969.696,37	33.703.843,65	
Lobsters	700.435,21	1.612.368,68	1.606.737,35	
Other crustaceans and molluscs	4.193.653,27	4.348.801,34	4.470.068,07	
Sub-total	*44.759.996,15*	*39.083.767,79*	*42.074.499,05*	
Agricultural Products and Derivatives				
Food Products	11.114.987,34	9.857.096,38	11.594.148,07	15.500.548,54
Cash crops	51.363.068,61	57.864.345,87	42.255.190,54	98.143.422,67
Products/ By-Products of Plant Origin	48.491.698,07	51.000.691,39	65.765.907,46	78.465.457,87
Products / By-products / Derived from Animal Origin	5.467.521,06	5.964.991,28	10.970.325,21	13.064.232,66
Sub-total	*116.437.275,08*	*124.687.124,92*	*130.585.571,29*	*205.173.661,74*
TOTAL	**161.197.271,23**	**163.770.892,71**	**172.660.070,33**	**205.173.661,74**
% of total exports			**0,68**	

Source: (MEF, 2014), Own illustration

Besides international trade, the economic value of biodiversity for Madagascar is also manifest in tourism industry and domestic trade. For example, according to World Bank data, number of tourists arrived in Madagascar just before the COVID-19 pandemic was 486,000 which was 35% increase compared to 2018 (World Bank, 2019). Most of these tourists are attracted to rich biodiversity of the country most famous of which include national parks of Ranomafana (host to bamboo lemurs), Isalo (host to diverse terrain and palm trees), Andasibe (several lemur species) and diverse marine natural habitats of whales among others (MEF, 2014).

Socio-cultural Values

Social values of biodiversity are listed in the latest National Report on Biodiversity (2014) as playing major role in food demand of people (especially, fishing, hunting and gathering in lean seasons), supply of building materials extensively used by local population (bamboo trees, coral beds, etc.), domestic fuels in form of woods, and traditional medicinal practices (use of flora and fauna species in healing/treating illnesses). On the cultural dimension, many strong ties exist between different local tribes' beliefs, lifestyles and specific flora and fauna species. For instance, several Malagasy tribes see lemurs and other animal species as their ancestors (hence they are forbidden for hunting), a lot of places have been given names of animals and plants, and sometimes a particular species is protected due its curses or because it is used as a place of sacred value (MEF, 2014). Because of diverse socio-cultural bonds between biodiversity elements and local population, some studies have suggested ways to improve coherence between conservation policies and culture (Fritz-Vietta N.V.M., Ferguson H.B., Stoll-Kleemann S., & Ganzhorn J.U., 2011).

Scientific Values

Scientific value of biodiversity for Madagascar is primarily linked to very high rates of endemism among its species and even family groups. Some species even reach 100% or near total endemism. These include but not limited to Aloe, palms, primates and the Orchidaceous (Rakotoarisoa, Ronell R. Klopper, & Gideon F. Smith , 2014; Franco Andreone, et al., 2008). These species represent a high degree of genetic diversity which opens up new and unexplored research fields for science to tap into and contributes to developments in genetic engineering for example (2014, S. 8).

Biodiversity Status

The current chapter briefly discusses the current status of biodiversity in Madagascar following the structure outlined in 5th National Report to Convention on Biodiversity; firstly, terrestrial and aquatic ecosystems are described narrowing down to the status of species and protected areas.

Ecosystem Status

The diverse habitats of Madagascar are categorized under terrestrial and aquatic ecosystems. These habitats are formed under climate, soil, altitude and human factors (MEF, 2014). **Terrestrial ecosystems** are, in turn, divided into forests, drylands and agricultural ecosystems according to their vegetation types.

Currently, the island has ten types of forests based on the altitude and climate characteristics of different ecoregions – from Tropical Rainforests and Western Dry Forests to Mangroves and Coastal Forests. Various facts regarding these forest types can be seen in Table 3. In addition, drylands are spread over a large territory, especially in the form of savannahs

(382.426 km2), but they account for only less than 5% of biodiversity and are mainly homogenous habitats. World Development Indicators of the World Bank depicts the latest agricultural land area as 70.3% of total land area or about 409.000 km2 which has remained more or less the same since 2001 (World Bank, 2018).

Aquatic ecosystems consist of a) wetlands and mainland waters and b) coastal and marine ecosystems. The former is spread over 5.339 km2 area and encompasses aquatic habitats such as lakes, swamps, rivers, streams and groundwater all of which originates from five major watershed areas of Madagascar – Montagne d'Ambre, Tsaratanana, the Eastern slope, the Western slope and the Southern slope (MEF, 2014). The coastal and marine ecosystems of Madagascar boast a rich biodiversity in forms of coral reefs, seagrass and algae and mangroves. Coral reefs extend over 2.230 km2 area of Madagascar shores (Sagrasswatch, 2021) and are found in the northwest, northeast and southwest coasts of the island with richness of respectively 318, 281 and 164 corals (MEF, 2014, S. 18). Seagrass and algae habitats are among least researched in Madagascar. However, conducted rapid assessments so far have confirmed 12 seagrass species and 5.796 km2 of seagrass meadow area in Madagascar north and south shores (Sagrasswatch, 2021). Lastly, the mangrove species of Madagascar are found to be 11 and mostly spread throughout the West coast covering 2.433 km2 area 16% of which is under protection (MEF, 2014).

Species Status

A bewildering spectrum of flora and wildlife reside in the island of Madagascar. Nearly 12.000 endemic flora species inclusive of seven baobab species (6 of which is found nowhere else), 202 palm tree species with 99% endemism rate, 1.100 orchid species with 86% endemism and numerous wild plants and alien, invasive species are distributed throughout the island together with broad array of ants (75% endemism), freshwater shrimps, crabs, fish (98 out of 115 being endemic and 29% them being of unknown status), amphibians, reptiles and mammals (WWF, 2021). Impressively, scientists continue to discover new wildlife species regularly. For instance, between 2010 and 2013, 19 new species of reptiles were added to the inventory which were unknown to science before (MEF, 2014).

Table 3. Forest ecosystems of Madagascar

	Type of forest ecosystems	Total area, km2	% protected area	Region
Climatic forest formation	Rainforest (tropical)	47.737	39	Eastern Ecoregion
	Degraded rainforest	58.058	2,45	
	Tapia forest	1.319	20,6	Central Ecoregion
	Rainforest (western)	72	-	Western Ecoregion
	Sub-humid forest	4.010	6,88	
	Dry forest	31.970	17,12	
	South-western spiny dry forest	18.355	4,46	Southern Ecoregion
	Degraded spiny dry forest	5.427	6,55	
Physiographic forest formation	Mountain dense rainforest			above 1600 m
	Silva Lichen			above 1800 m
Edaphic (soil) forest formation	Mangroves	2.433		coastal rainforest zone
	Coastal forest	274	13,83	
	Swamp forest			
	Riparian forest			Along the rivers

Source: (MEF, 2014), Own illustration

Trends in Biodiversity

Trends at Ecosystems Level

Perhaps one of the most alarming trends regarding ecosystem trends in Madagascar is the rate at which annual deforestation happened over past decades and still continues. A recent paper published in Biological Conservation journal (Vieilledent, et al., 2018), analysed deforestation in Madagascar throughout last six decades (1953-2014) by combining forest cover maps and annual tree loss datasets. The results found indicated that the Big Island has lost 44% of its forests during the examined period with deforestation rates rising up to 1.1%/year (or 99000 ha/year) in the most recent years. The data provided in the 5th National Biodiversity Report provides more detailed information on deforestation during 2005-2010 based on the altitude of natural forests (MEF, 2014). It reports that "because of their easier accessibility, natural forests of low altitude (0-400m) underwent a higher deforestation rate, 0.5% per year" while the forests located at higher altitudes and inside the protected areas went through a comparably lower rates of loss in that period.

The similar negative trends are observed wetland, marine and coastal ecosystems of the country. Deforestation, wetland siltation and expansion of rice fields are main factors behind drying and, in some cases, disappearing lakes, marshlands and this affects the endemic species living in these ecosystems in the most harmful ways (MEF, 2014). Meanwhile, the coral reef cover of Northwest and Northeast Madagascar are reported be in in the good shape, but the serious degradation is observed in Toliara Grand Recif in terms of architectural species habitat being invaded by algae (MEF, 2014, S. 40; Raymond, 2020). Main causes of mangrove and coral reef deteriorations are found to be ocean warming and acidification due to climate change as well as overexploitation and increased pollution due to population growth (Raymond, 2020).

Trends at Species Level

The latest and first comprehensive status evaluation report indicates that out of 3.118 tree species in Madagascar 93% is endemic to the island and 1.828 of these (63%) are threatened seriously (from critically endangered status to vulnerable) (Emily Beech, et al., 2021). Palm trees are the most at risk of extinction plant groups in Madagascar with 83% of its species under extinction threat (MEF, 2014).

Trend in wildlife groups, species and populations seem to go in the negative direction of biodiversity loss according to 2014 National Report on Biodiversity (MEF, 2014). Although the newly uncovered amphibian and reptile species is increasing (16.8% and 5.14% in 2014), population size and distribution range of some species are decreasing classifying them as critically endangered and in extinction threat such as green turtles, olive Ridley and Loggerhead turtles.

Many of the species are dependent on the forest habitats and therefore are vulnerable to deforestation trends. Among such fauna species are the lemurs whose status has altered to critically endangered and vulnerable for different species in recent years due to decline in population sizes. In addition, five new species of lemurs were uncovered between 2009 and 2014 and even new populations discovered for *Prolemur simus* (MEF, 2014).

Threats to Biodiversity

In general, the biological diversity in Madagascar is among the most threatened in the world because of various development factors of different nature. The flora and fauna of the fourth biggest island in the world are threatened by overexploitation, change of land use patterns, industrial development, pollution, climate change, abundance of invasive species, fires, deforestation and diseases not only at species level but even at family and ecosystem levels (MEF, 2014).

The latest national report (MEF, 2014) indicates that some plant and animal species exploited mostly for selling in local markets such as Aloe, Orchid species, Eugenia used as ornamental plants and in local alcoholic drink manufacturing and overfishing in especially North-west region of the island. Among large-scale industries, logging for mainly charcoal and fuelwoods is the most excessive practice destroying natural habitats of rosewood and ebony trees. According a FAO report, in 2010, 98% of all round wood logged in all Madagascar forests was for charcoal production (S. 47). Other irrational exploitation practices include collection of bird eggs, catching of whales, dolphins and hunting exotic wildlife species such as lemurs either for local business or exporting.

Unsustainable land use practices are mostly driven by agricultural purposes. Lake watersheds are reduced due to logging and land clearing (e.g. burning) for increased rice cultivation. Even simple practices such as installing agricultural drainage materials in soil can alter and risk extinction of land-specific structures and features such as in Torotorofotsy Ramsar Site (MEF, 2014, S. 48).

Recent discoveries of precious metals and stones like gold, as well as existing oil and mining industries are of very serious consequences to natural habitat species of some vulnerable areas. One recent example is what is known as 'sapphire rush' in Eastern rainforests of Madagascar, Diddy area where locals excessively mined for precious sapphires and rubies to sell in local markets and this practice put under question the survival of precious species living in those forests such as Indri lemurs (Jones, 2016).

Another source of threat that is more related to population access to sanitation and daily practices is pollution of aquatic ecosystems (MEF, 2014). USAID Water and Development Country Plan for Madagascar indicates that "44 percent of Madagascar's population still practices open defecation" (USAID, 2020, S. 2). This underdevelopment combined with lack of awareness regarding the consequences of water pollution poses serious dangers to watersheds, wetlands and marine ecosystems of the island.

An indirect consequence of anthropogenic intrusion is proliferation of invasive species that sometimes can be highly threatening for native flora and fauna. These invasive organisms are dangerous for local habitats since they are very resilient and adaptive to different conditions, thus they may spread in expense of areas and nutrients needed for survival of other species. In Madagascar, for example, Northwest coasts has seen occurrences of abundance in green algae provoked by overfishing which consequently brought about coral degradation and underdevelopment of spiny sea urchins (MEF, 2014, S. 49).

Climate change impacts is yet another and one of the most serious long-term threats that the Madagascar biodiversity faces. In terms of marine habitats, rise of water surface temperatures has significant impact on coral reefs and mangroves of Southwest (MEF, 2014). The same factors are expected to have impacts on breeding and distribution of sea turtles and amphibian species as well. Furthermore, a recent scientific paper published in Plant Diversity journal (Wan, et al., 2021) found that climate change is expected to both expand and contract suitable habitats for all endemic baobab species of Madagascar, while the original habitat loss is forecasted to be large-scale especially for the specie of Adansonia za (40% loss of original habitat predicted due to climate change).

The problem of deforestation is a direct threat to many ecosystems in Madagascar including mangroves and coral reefs of marine and wetland habitats, as well as terrestrial wildlife. A study conducted published in 2020 used the genus Varecia or ruffed lemurs as an indicator of the eastern rainforest biodiversity status and found that by 2070 even if further deforestation is prevented by effective measures of protected area systems, the climate change alone would cause a 62% loss of appropriate habitat for ruffed lemurs (Morelli, et al., 2020). The same study also calculated a loss of habitat of 81% adding the ongoing deforestation in eastern rainforests of Madagascar. Therefore, the current status of biodiversity of both rainforests and other endangered ecosystems requires an effective maintenance and integrity enhancement of protected area networks in Madagascar.

Nature Conservation Programs

In the National Biodiversity and Action Plans 2015-2025, Madagascar's vision for biodiversity conservation is stated as: "A Malagasy people living in harmony in a country where the environment is preserved and / or restored, taking advantage for its welfare and sustainable use and enhancement and reasoned with rich and valued biodiversity, resilient to environmental change" (Harison, Sahoby Ivy, Fara Mihanta, & Hanitra Lalaina, 2016, S. 66). To achieve this vision, several national and international tools such as nature conservation programs are in place. The objective of these measures is not only improving status of biodiversity, but also linking it to overarching goals of well-being and poverty reduction (2016, S. 67).

Madagascar, in 5th World Parks Congress of 2003, took an ambitious responsibility to increase the coverage of protected areas from 1.7 million ha in that year to 6 million ha in order to conserve the country's rich biodiversity (Durbin, 2007). Many stakeholders including the Government, funding organizations, NGOs, private companies and individuals collaborate in order to create, expand and sustainably manage the protected areas which fall under five of seven categories defined by International Union for Conservation of Nature (IUCN) (Table 4).

Table 4. Madagascar protected areas according to IUCN categories

IUCN categories	***Explanation***	***Number in Madagascar***	***% in Madagascar***
Ia. Strict Nature Reserve	all-inclusive (species and geological features) ecosystems that are protected strictly	2	1.17%
II. National Park	Also provide recreational services to the public	24	14.04%
IV. Habitat or Species Management Area	Purpose is to protect particular ecosystems and species and prioritize them in the management	22	12.87%
V. Protected Landscape or Seascape	purpose is to maintain balanced interaction between humans and nature	24	14.04%
VI. Protected area with sustainable use of natural resources	"low-level non-industrial use of natural resources compatible with nature conservation" is allowed in these areas (IUCN, 2021)	4	2.34%
Not Reported	-	90	52.63%
Not Applicable	-	5	2.92%

Source: (IUCN, 2021; UNEP-WCMC and IUCN, 2021), Own illustration

According to the online database Protected Planet (UNEP-WCMC and IUCN, 2021), Madagascar currently has a network of 171 protected areas (PA) covering an overall area of more than 55.000 km2 (see Figure 1 for growth pattern of PAs in Madagascar). Out of this, terrestrial protected area coverage is 7,5%, whereas marine protected area coverage is slightly below 1%. According to the database, the PAs are managed and governed in a mixed way; most of them by local communities (81), and very little of them by government-delegated management (2), federal or national ministry or agency (1), non-governmental organizations (2) and so on. This mixed management practices presents its own effectiveness challenges (Gardner, et al., 2018).

There are many projects supported by international donor organizations that funds or further facilitates the implementation and management of new system of protected areas in Madagascar. One example includes Managed Resources Protected Areas (MRPA) project with UNDP and GEM, which directly supports five sites: Mahavavy Kinkony Complex, Ambohimirahavavy Marivorahona, Ampasindava Galoko-Kalobinono, Menabe Antimena and Loky Manambato (MEF, 2014). Several conservation initiatives especially aimed at protection of lemurs in Madagascar have been taken by Duke Lemur Center (Duke Lemur Center, 2021), Biodiversity Conservation Madagascar (Lemur Conservation Network, 2021) and others. The Duke Lemur Center's nature conservation projects include DLC-SAVA, ex-situ lemur Initiatives, Conservation Breeding programs among others. The DLC-SAVA program focuses on education of people in Madagascar to have a positive long-term impact on biodiversity conservation and sustainable resource use and management. On the other hand, ex-situ lemur initiatives contain animal husbandry, breeding and animal welfare programs for lemurs living in zoos and other conservation facilities and supported by the Government of Madagascar. There are also some volunteer organizations such as Podvolunteer whose mission is to help protect and conserve tropical forest ecosystems of Madagascar coastal areas through scientific research on biodiversity and encouraging sustainable community actions (Podvolunteer, 2021).

Figure 1. Growth in Protected Area Coverage, Source: (UNEP-WCMC and IUCN, 2021)

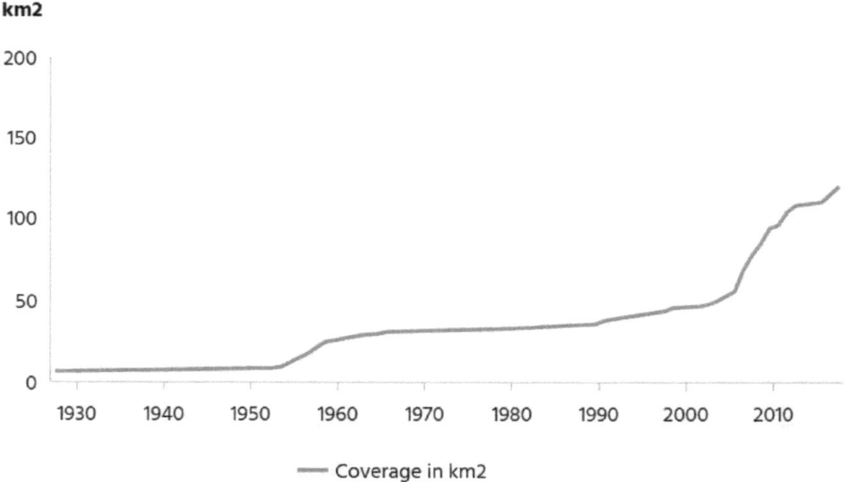

Section Conclusion

Madagascar is the party by ratification to the CBD since 1996 and to its two protocols (Cartagena and Nagoya) since the beginning of 21st century (Convention on Biological Diversity, 2021). Under this framework, the country has put in place different legal, economic and partnership tools in order to achieve its targets of preserving vast richness of the flora and fauna as well as different ecosystems and habitats that are home to dazzling amount and variety of endemic species. Some improvements have been achieved in terms of preventing the loss of biodiversity and enabling a healthier environment for some species. However, the latest reports show that there are alarmingly negative recent and long-term trends such as deforestation, overexploitation and decrease in population size, range and species number of individual plant and animals. These trends are reinforced by factors of climate change and subsequent increased droughts, poverty and subsistence farming. Scientific researchers find that endemic species are the most vulnerable ones to above-mentioned trends.

Considering the economic, socio-cultural, scientific and other values of biodiversity richness to Madagascar, it is critical to enhance existing programs, especially the system of protected areas, and educate and raise awareness of local population on the importance of nature conservation. These steps must continue to focus on poverty reduction and national wellbeing and integrate policies and actions at different sectors and levels throughout the country, region and international level.

Ultimately, the sustainable development of any country is dependent upon the harmony and balanced relationships between nature (ecosystems, species, landscapes, etc.) and human beings. Therefore, biodiversity is an agenda that is always relevant and important to the global wellbeing.

References

Tanzania

Biodiversity in Tanzania. (n.d.). Tanzanian Clearing House Mechanism. https://tz.chm-cbd.net/biodiversity (accessed 03.07.2021)

IUCN (2017) Governance of protected and conserved areas in Tanzania. Phase 1 workshop as part of an IUCN-assisted process of assessment and action to enhance governance for conservation and sustainable livelihoods. Dar es Salaam

Kideghesho JR, Rija AA, Mwamende KA, Selemani IS (2013) Emerging issues and challenges in conservation of biodiversity in the rangelands of Tanzania. Nature Conservation 6: 1–29. doi: 10.3897/natureconservation.6.5407

Latest NBSAP´s Tanzania. (n.d.). Convention on Biological Diversity. https://www.cbd.int/nbsap/about/latest/#tz (accessed 03.07.2021)

Mascarenhas, A. C., Ingham, Kenneth, Chiteji,. Frank Matthew and Bryceson,. Deborah Fahy (2021). Tanzania. Encyclopedia Britannica. https://www.britannica.com/place/Tanzania (accessed 03.07.2021)

Millennium Ecosystem Assessment, (2005). Ecosystems and Human Well-being: Synthesis. Island Press, Washington, DC.

Newmark, William D. (2002) What is biodiversity, University of Utah. Doi: 10.1007/978-3-662-04872-6_1

Promoting Tanzania's Environment, Conservation, and Tourism through the USAID PROTECT Activity. (n.d.). RTI International. https://www.rti.org/impact/tanzania-conservation-project (accessed 03.07.2021)

Tanzania. (n.d.). National Geographic. https://www.nationalgeographic.co.uk/topic/locations/earth/africa/tanzania (accessed 03.07.2021)

Torell, Elin, Mwanahija Shalli, Julius Francis, Baraka Kalangahe, Renalda Munubi (2007) Tanzania Biodiversity Threats Assessment: Biodiversity Threats and Management Opportunities for

Fumba, Bagamoyo, and Mkuranga, Coastal Resources Center, University of Rhode Island, Narragansett, 47 pp.

United Republic of Tanzania - Main Details. (n.d.). Convention on Biological Diversity. https://www.cbd.int/countries/profile/?country=tz (accessed 03.07.2021)

United Republic of Tanzania (2014) Fifth National Report on the Implementation of the Convention on Biological Diversity. Vice President's Office, Division on Environment, Dar es Salaam

United Republic of Tanzania (2015) National Biodiversity Strategy and Action Plan (NBSAP) 2015-2020. Vice President's Office, Division on Environment, Dar es Salaam

United Republic of Tanzania (2019) Sixth National Report for the Convention on Biological Diversity. Vice President's Office, Division on Environment, Dar es Salaam

United Nations Environment Programme (2021) How the world can head off a looming biodiversity crisis. https://www.unep.org/news-and-stories/story/how-world-can-head-looming-biodiversity-crisis (accessed 03.07.2021)

WCS Tanzania. (2016). WCS Tanzania Program. https://tanzania.wcs.org/ (accessed 03.07.2021)

Madagascar

Burton V. Barnes, & Marc Lapin. (1995, October). Using the Landscape Ecosystem Approach to Assess Species and Ecosystem Diversity. *Conservation Biology, 9*(5), 1148-1158. doi:10.1046/j.1523-1739.1995.9051134.x-i1

Convention on Biological Diversity. (1992). *Article 2. Use of Terms*. Retrieved from Convetion on Bilogical Diversity: https://www.cbd.int/convention/articles/?a=cbd-02

Convention on Biological Diversity. (2021). *Madagascar - Main Details*. Retrieved from Convention on Biological Diversity: https://www.cbd.int/countries/profile/?country=mg

Duke Lemur Center. (2021). *OVERVIEW: MADAGASCAR CONSERVATION PROGRAMS*. Retrieved July 03, 2021, from Duke Lemur Center: https://lemur.duke.edu/protect/overview-madagascar-conservation-programs/

Durbin, J. (2007, February 02). Madagascar's new system of protected areas – Implementing the 'Durban Vision' . Retrieved July 03, 2021, from https://conservation-development.net/Projekte/Nachhaltigkeit/CD2/Madagaskar/Links/PDF/01DurbanVision.pdf

Emily Beech, Malin Rivers, Marina Rabarimanarivo, Noro Ravololomanana, Nadiah Manjato, Faranirina Lantoarisoa, . . . Vololoniaina Jeannoda. (2021). *The Red List of Trees of Madagascar.* Botanic Gardens Conservation International. Retrieved from file:///C:/Users/mreel/Downloads/MadagascarMedRes.pdf

Explore the Biodiversity Hotspots. (2021). Retrieved July 01, 2021, from Critical Ecosystem Partnership Fund: https://www.cepf.net/our-work/biodiversity-hotspots

Fritz-Vietta N.V.M., Ferguson H.B., Stoll-Kleemann S., & Ganzhorn J.U. (2011). Conservation in a Biodiversity Hotspot: Insights from Cultural and Community Perspectives in Madagascar. (Z. F., & Habel J., Eds.) *Biodiversity Hotspots*, 209-233. doi:https://doi.org/10.1007/978-3-642-20992-5_12

Gardner, C. J., Nicoll, Martin E., Rakotomalala, Christopher Birkinshawc, Alasdair Harris, Richard E. Lewis, . . . Anitry N. Ratsifandrihamanana. (2018). The rapid expansion of Madagascar's protected area system. *Biological Conservation, 220,* 29-36. doi:https://doi.org/10.1016/j.biocon.2018.02.011

Harison, R., Sahoby Ivy, R., Fara Mihanta, A., & Hanitra Lalaina, R. (2016). *National Biodiversity and Action Plan 2015-2025.* Republic of Madagascar. Retrieved from https://www.cbd.int/doc/world/mg/mg-nbsap-v2-en.pdf

Hughes, A. R., Brian D.Inouye, Marc T. J. Johnson, Nora Underwood, & Mark Vellend. (2008). Ecological consequences of genetic diversity. (O. Schmitz, Ed.) *Ecology Letters*(11), 609-623. doi:10.1111/j.1461-0248.2008.01179.x

IUCN. (2021). *Protected Area Categories.* Retrieved from IUCN: https://www.iucn.org/theme/protected-areas/about/protected-area-categories

Jones, J. P. (2016, November 21). *A 'sapphire rush' has sent at least 45,000 miners into Madagascar's protected rainforests.* Retrieved July 02, 2021, from The Conversation: https://theconversation.com/a-sapphire-rush-has-sent-at-least-45-000-miners-into-madagascars-protected-rainforests-69164

Lemur Conservation Network. (2021). *Biodiversity Conservation Madagascar.* Retrieved from Lemur Conservation Network: https://www.lemurconservationnetwork.org/organization/biodiversity-conservation-madagascar/

Macarthur, R. H. (1965, November). Pattersn of Species Diversity. *Biological Reviews, 40*(4), 510-533. doi: https://doi.org/10.1111/j.1469-185X.1965.tb00815.x

MEF. (2014). *The 5th National Report to the Convention on Biological Diversity.* Ministry of Environment. Ministry of Environment. Retrieved from https://www.cbd.int/doc/world/mg/mg-nr-05-en.pdf

Morelli, T. L., Adam B. Smith, Amanda N. Mancini, Elizabeth A. Balko, Cortni Borgerson, Rainer Dolch, . . . Andrea L. Baden. (2020). The fate of Madagascar's rainforest habitat. *Nature Climate Change, 10*, 89-96. doi:https://doi.org/10.1038/s41558-019-0647-x

Podvolunteer. (2021). *Conservation Madagascar.* Retrieved July 03, 2021, from podvolunteer: https://www.podvolunteer.org/projects/conservation-madagascar

Raymond, E. (2020). *Annex F: Climate Change Analysis.* CDC, USAID. Retrieved July 03, 2021, from https://www.climatelinks.org/sites/default/files/asset/document/2021-03/2021_USAID_CDC%20Annex-Madagascar.pdf

Rockström, J., Will Steffen, Jonathan A. Foley , Kevin Noone, Åsa Persson, F. Stuart Chapin III, . . . Paul Crutzen. (2009, September 23). A safe operating space for humanity. *Nature, 461*, 472-475. doi:https://doi.org/10.1038/461472a

Sagrasswatch. (2021). *Madagascar.* Retrieved July 02, 2021, from Seagrass-Watch: https://www.seagrasswatch.org/madagascar/#footnote

UNEP-WCMC and IUCN. (2021, July). Protected Planet: The World Database on Protected Areas (WDPA) and World Database on Other Effective Area-based Conservation Measures (WD-OECM). Cambridge, UK. Retrieved from www.protectedplanet.net

USAID. (2020). *Water and Development Country Plan.* USAID. Retrieved July 02, 2021, from https://www.globalwaters.org/sites/default/files/madagascar_country_plan_2020.pdf

Vieilledent, G., Clovis Grinand, Fety A. Rakotomalala, Rija Ranaivosoa, Jean-Roger Rakotoarijaona, Thomas F. Allnutt, & Frédéric Acharda. (2018, June). Combining global tree cover loss data with historical national forest cover maps to look at six decades of deforestation and forest fragmentation in Madagascar. *Biological Conservation, 222,* 189-197. doi:https://doi.org/10.1016/j.biocon.2018.04.008

Wan, J.-N., Ndungu J.Mbari, Sheng-WeiWang, Brian N.Mwangi, Jean R.E. Rasoarahona, Bing Liu, . . . Qing-Feng Wang. (2021, April). Modeling impacts of climate change on the potential distribution of six endemic baobab species in Madagascar. *Plant Diversity, 43*(2), 117-124. doi:https://doi.org/10.1016/j.pld.2020.07.001

World Bank. (2018). Agricultural land (sq. km) - Madagascar. World Bank. Retrieved July 01, 2021, from https://data.worldbank.org/indicator/AG.LND.AGRI.K2?locations=MG

World Bank. (2019). International tourism, number of arrivals - Madagascar. *World Development Indicators*. World Bank. Retrieved July 01, 2021, from https://data.worldbank.org/indicator/ST.INT.ARVL?end=2019&locations=MG&start=1995&view=chart

WWF. (2021). *Madagascar*. Retrieved from World Wildlife: https://www.worldwildlife.org/places/madagascar